An Animal Family

GOATS AND KIDS

By Natalie Humphrey

Gareth Stevens
PUBLISHING

Please visit our website, www.garethstevens.com. For a free color catalog of all our high-quality books, call toll free 1-800-542-2595 or fax 1-877-542-2596.

Library of Congress Cataloging-in-Publication Data

Names: Humphrey, Natalie, author.
Title: Goats and kids / Natalie Humphrey.
Description: New York : Gareth Stevens Publishing, [2021] | Series: An animal family | Includes index.
Identifiers: LCCN 2019045312 | ISBN 9781538255728 (library binding) | ISBN 9781538255704 (paperback) | ISBN 9781538255711 (6 Pack) | ISBN 9781538255735 (ebook)
Subjects: LCSH: Goats–Infancy–Juvenile literature. | Goats–Development–Juvenile literature.
Classification: LCC SF383.35 H86 2021 | DDC 636.3/9–dc23
LC record available at https://lccn.loc.gov/2019045312

First Edition

Published in 2021 by
Gareth Stevens Publishing
111 East 14th Street, Suite 349
New York, NY 10003

Editor: Natalie Humphrey
Designer: Katelyn E. Reynolds

Photo credits: Cover, p. 1 CyberKat/Shutterstock.com; pp. 5, 24 (herd) Simon Werner/500px Prime/Getty Images; p. 7 Amir Mukhtar/Moment/Getty Images; p. 9 schubbel/Shutterstock.com; p. 11 DonnaA Country Photos/Shutterstock.com; p. 13 vasiliki/E+/Getty Images; p. 15 Malachi Jacobs/Shutterstock.com; p. 17 Traci Law/Shutterstock.com; p. 19 Nina Skarga/Shutterstock.com; pp. 21 (left), 24 (mountain) Josh Schutz/Shutterstock.com; pp. 21 (right), 24 (farm) magesbycat/Shutterstock.com; p. 23 imagevixen/Shutterstock.com.

Printed in the United States of America

CPSIA compliance information: Batch #CS20GS: For further information contact Gareth Stevens, New York, New York at 1-800-542-2595.

Contents

A herd is a goat family.

Herds can have 30 goats.

A baby goat is
called a kid.

Mothers have two kids a year.

Mother goats make milk.

Kids need milk to grow.

They need milk for
six weeks.
Some need it longer.

Mothers teach kids what to do.

Some goats live
on mountains.
Some live on farms.

Kids are very friendly.
They play with
their herd.

Words to Know

farm

herd

mountain

Index